The Truth about
Dangerous SEA CREATURES

Written by Mary M. Cerullo   Photographs by Jeffrey L. Rotman   Illustrations by Michael Wertz

chronicle books · san francisco

Dedicated to my brother Tom —M. M. C.
To Mary, who has given my photographs the gift of speech —J. L. R.

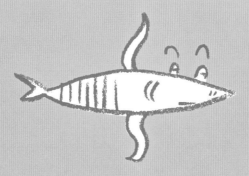

Thanks to Dot Wensick, librarian at the New England Aquarium, and Laura Burkhart,
reference librarian at the California Academy of Sciences; to Brian Nelson, senior aquarist,
New England Aquarium, for answering my questions and reviewing the manuscript;
also to Bill Curtsinger and Jeff Rotman for sharing their firsthand experiences
with many of these animals. —M. M. C.

Text © 2003 by Mary M. Cerullo.
Photographs © 2003 by Jeffrey L. Rotman.
Illustrations © 2003 by Michael Wertz.
Photograph on page 16 (Portuguese man-of-war) © David Schrichte/Seapics.com

Book design by Jessica Dacher.
Typeset in Centaur and Franklin Gothic.
Manufactured in China.

Library of Congress Cataloging-in-Publication Data
Cerullo, Mary M.
The truth about dangerous sea creatures / by Mary M. Cerullo ;
photographs by Jeffrey L. Rotman ; illustrated by Michael Wertz.
p. cm.
Includes bibliographical references (p. 45). Contents: Giant squid-Giant octopus-Blue-ringed
octopus-Giant clam-Spiny sea urchin-Crown-of-thorns sea star-Jellies-Sea wasp-Surgeonfish-
Cone shell-Stonefish-Lionfish-Coral reefs-Damselfish-Sea anemone-Sea snake-Barracuda-
Giant grouper-Puffer fish-Manta ray-Stingray-Torpedo ray-Basking shark and whale shark-
Tiger shark-Bull shark-Great white shark.
ISBN 0-8118-4050-6
1. Dangerous marine animals-Juvenile literature. [1. Dangerous marine animals.
2. Marine animals.] I. Rotman, Jeffrey L., ill. II. Wertz, Michael., ill. III. Title.
QL100.C47 2003
591.77—dc21
2003000828

Distributed in Canada by Raincoast Books
9050 Shaughnessy Street, Vancouver, British Columbia V6P 6E5

10 9 8 7 6 5 4 3 2 1

Chronicle Books LLC
85 Second Street, San Francisco, California 94105

www.chroniclekids.com

Contents

What is it that makes us think twice about jumping into a sparkling blue ocean?

Are we letting our imaginations get the better of us? Have we seen one too many horror movies about killer sharks and giant squid? Or do we feel like trespassers in an alien world where we don't understand the customs and behaviors of its inhabitants?

Sea creatures have an arsenal of weapons that a ninja warrior would envy:

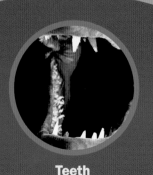

Teeth

Size

Spines

Speed

Daggers

We don't need to feel unwelcome in the world beneath the sea if we simply remember two important rules. The first governs the residents, and the other should govern us, their guests:

Eat or be eaten.
Look but don't touch.

Once we get to know the creatures of the ocean, we can decide for ourselves whether or not they deserve their fearsome reputations.

Surprise

Poison

Strength

Camouflage

Electricity

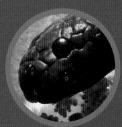

Venom

HELP!

The Truth Is ...

This Sea Monster DOES Exist—We've Just Never Seen One Alive

Giant Squid

On old nautical charts, drawings of sea monsters decorated the unexplored regions of the ocean. These maps often featured the giant kraken with its snakelike arms wrapped around the tall masts of a sailing ship. According to legend, it could pluck a sailor off the deck or pull a ship to the bottom of the sea.

For centuries people believed these were just sea stories, that is, until the remains of several giant squid washed ashore in the late 1800s. Others were found tangled in fishing nets, dying at the surface of the sea, and some were discovered inside the stomachs of sperm whales, their chief enemies. So far, no one has ever seen a healthy, living specimen.

Not long ago, Dr. Clyde Roper, a scientist at the Smithsonian Museum in Washington, D.C., descended into the deep ocean off New Zealand in search of a giant squid in its natural habitat. The animal Dr. Roper was looking for has eyes the size of the portholes in his minisub. With its tentacles outstretched, it may measure 60 feet (18 meters) long. Its many suckers are rimmed with small hooks that dig into the flesh of its prey. Inside its mouth is a parrotlike beak that can bite through steel cable. Unlike most of us would have been, Dr. Roper was disappointed that he didn't find his giant squid.

The Most Dangerous Octopus Is One of the Smallest and Prettiest

Blue-ringed Octopus

Beautiful but deadly, the blue-ringed octopus of the South Pacific is only 6 inches (15 centimeters) long. The blue-ringed octopus often creeps across tide pools at night, hunting for crabs, snails, and small fish. If a human were to pick up the tiny octopus, its blue spots would darken in alarm, and it would nip its "attacker" and spit powerful venom into the wound. At first, the person might hardly notice the bite, but within a few minutes he or she would start to feel the effects of the venom, called *tetrodotoxin*, which is not made by the octopus itself but by several kinds of bacteria that live inside its body. The venom is sometimes strong enough to kill a human within minutes.

Giant Octopus

Imagine going through a growth spurt where you grow 6 inches (15 centimeters) in a week! That's what the giant octopus, which lives for only three years, must do to reach 15 feet (4.6 meters) from arm tip to arm tip. Despite its size, the giant octopus doesn't attack humans, but if you stick your hand inside an octopus's den, you may find yourself gripped by an overwhelming force. Each of its suction cups holds on tightly by creating a vacuum seal inside the sucker—much like the lid on a jar of homemade jelly.

Jim Cosgrove is a Canadian biologist who has danced with giant octopuses. Dr. Cosgrove has been diving among giant octopuses in the cold waters of the North Pacific Ocean for over 20 years. More than once, a curious octopus has draped itself around Jim's shoulders and spun him around a few times before letting go. Says Dr. Cosgrove, "An octopus explores you and envelops you. It doesn't try to bite. It's actually enjoyable."

1, 2, CHA CHA CHA!

The Giant Clam Is a Killer Only in Hollywood

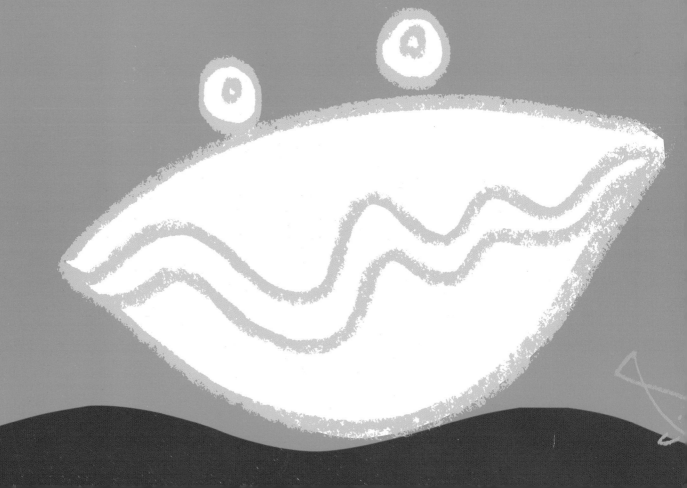

Giant Clam

In old-time adventure movies, one of the underwater perils that the sunken treasure-hunting hero had to face was the steam-shovel-like jaws of a giant clam, which would threaten to clamp its shells together and trap the diver at the bottom of a tropical sea. In real life, a diver would have to be awfully slow-moving or slow-witted to get his or her foot wedged between the shells of a giant *Tridacna* clam. Not that these clams aren't big; they can grow to be 4 feet (1.2 meters) across.

The clam closes its shells together as soon as it feels a touch against its mantle, the soft tissue lining the inside of its shells. A strong muscle at the base of the two shells (the same muscle we eat from scallops) keeps the shells clenched together until it feels no more movement. If you did get yourself stuck inside a giant clam, you could cut this muscle and free yourself. Treasure-hunting heroes might also be reassured to know that fully grown giant clams can't close their shells completely.

Could this giant lobster be another Hollywood movie monster?

Most Divers Fear This Animal More Than Any Other

Spiny Sea Urchin

One of the most common diving injuries comes from bumping into this tropical sea urchin. Its long, brittle spines break off easily when touched, leaving portions in your hand or foot. Then when you try to pull them out, they can break again, leaving small bits of spine inside the wound. The embedded spines take weeks or months to completely dissolve under your skin. Some sea urchins' spines are filled with venom and some are not, but a wound from either can easily become infected.

Spiny sea urchins hide in coral reefs by day. At night they march spine-to-spine in a triangular formation across the sandy bottom to feed on sea grass and seaweed.

Crown-of-Thorns Sea Star

Like the sea urchin, the crown-of-thorns sea star has spines all over its body. Its spines, though shorter than the sea urchin's, can also puncture the skin of a person's hand or foot.

Worst of all, the crown-of-thorns attacks living coral. It can grow to be 2 feet (60 centimeters) across and have 12 or more arms. Hordes of these sea stars have devoured large sections of the Great Barrier Reef of Australia, leaving only white coral skeletons behind.

When Is a Fish Not a Fish?

When it's a starfish or a jellyfish. A fish has a backbone (it's a *vertebrate*), as well as fins, gills, and scales. Starfish and jellyfish are animals without backbones (called *invertebrates*). Scientists now use the terms *sea stars* and *jellies* to help people avoid confusing these creatures with what they aren't—fish.

~~Jellyfish~~
Jelly

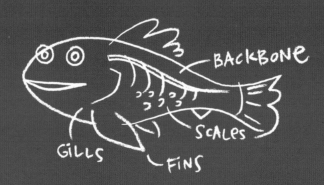

BACKBONE
SCALES
GILLS
FINS

A real fish!

~~Starfish~~
Sea star

Swimmers can encounter the stinging tentacles of the Portuguese man-of-war as far as 30 feet (9 meters) from its bright blue bell.

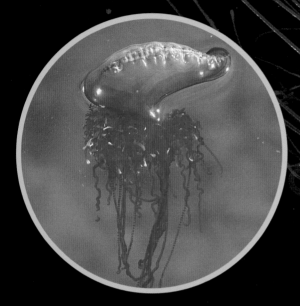

The Truth Is...

Some Animals Are As Dangerous Dead As Alive

Jellies

What has no head, no spine, and no heart and is 95 percent water? A jellyfish. Actually, scientists now call them *jellies* because they aren't fish at all, but cousins of sea anemones and corals. There are many different kinds of jellies, some as small as a pea and others up to 7 feet (2.1 meters) wide. What most jellies have in common is a body shaped like an umbrella that opens and closes to propel them gracefully through the water.

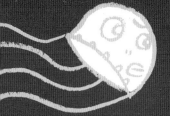

jellies!

Although jellies may look delicate, they can be powerful killers. Jellies capture fish using stinging cells that shoot tiny, poison-tipped harpoons into their victims. The venom paralyzes the prey while long tentacles pull the victim into an opening inside the umbrella, which is the closest thing a jelly has to a mouth. Human reactions to a jelly's sting range from a mild ache to convulsions and death. Even a dead jelly can hurt you because its stinging cells are released automatically whenever they are touched. Some jellies, such as the Portuguese man-of-war, can inflict painful welts that leave human victims writhing. Others, such as moon jellies, leave no impression on humans at all.

Sea Wasp

The most dangerous jelly of them all is the sea wasp. It's also called a box jelly because of its square shape. A fast swimmer, it pulses powerfully through the water while other jellies are content to drift gently with the current. Sometimes the sea wasp comes into the shallow coastal waters of Australia in search of food, which can put it in contact with humans. Curiously, its poison darts are triggered only by the chemicals in living things, such as fish, shrimp, and humans. It doesn't waste its ammunition on things that aren't potential food.

A collision with a sea wasp can kill a swimmer within a few minutes. One sea wasp, which is usually the size of a grapefruit, has enough venom to kill 60 adults.

Most Dangerous Marine Animals Live in the Tropics

Coral Reefs

Why are tropical waters home to most of the dangerous animals in the ocean? Scientists (and the people who live around the equator) have wondered that, too. Coral reefs are some of the oldest natural communities on Earth. The first modern reef-building corals date back 200 million years, so reefs have had a long time to develop a wide variety of life-forms.

A single Caribbean coral reef may be home to more than a thousand different kinds of plants and animals. Except for the rain forest, a coral reef is the world's most complex ecosystem. Coral reefs have been described as cities beneath the sea. They teem with animals and plants living at different levels of the reef, like apartment dwellers in high-rise skyscrapers. The creatures of the reef have evolved special characteristics and habits that help them compete for limited space and food supply. Some of those characteristics can be dangerous to humans.

Sea Anemone

Other reef creatures have set up unusual alliances. This partnership is called *symbiosis,* which means "living together." The clownfish, for example, lives inside the stinging tentacles of a sea anemone, an animal rather like an upside-down jelly. Like a jelly, the sea anemone has venomous stinging cells to kill its prey. The clownfish snuggles up against the anemone so that some of the anemone's mucous coating rubs off on it. This evidently leads the anemone to believe the fish is an extension of itself rather than an outsider. The clownfish gains a hiding place, and in exchange it lures other fish into the stinging tentacles of the sea anemone. Even a human wouldn't try to break up this partnership because tropical sea anemone stings can cause a painful rash.

Some Creatures Have Hidden Weapons

Cone Shell

There are over 400 different kinds of cone shells, some so beautifully patterned that shell collectors pay thousands of dollars for a single specimen. Some of the snails that live in these kinds of cone shells, however, contain a nasty surprise for those greedy enough to scoop them away from their tropical home. When disturbed, those snails immediately shoot a poison dart. If this dart strikes the hand of an undersea collector, he or she will probably feel a burning sting, followed by numbness and a tingling sensation that spreads over the whole body. In extreme cases, a human can become paralyzed and fall into a coma.

The cone shell prefers to use its venom to catch a meal, such as worms and very small fish. It may shoot several darts into its victim and then drag the paralyzed prey toward its shell. The opening of its shell is too small to swallow most animals, so the cone shell usually has to digest its victim outside its shell. It engulfs the animal with an expandable "mouth."

Stonefish

Still as a stone, the stonefish lies motionless on the ocean floor. Drab coloring, a lumpy body, and ragged skin that looks like seaweed complete the disguise. The stonefish pounces on small fish and crabs that stray too close to its large mouth. Considering how well camouflaged it is against predators, it's a mystery why the stonefish is also the most poisonous fish in the world. It has spines along its back and side fins that are like poison-tipped hypodermic needles. Bumping into this animal can be a fatal mistake for a human swimmer.

Surgeonfish

Spines as sharp as scalpels (surgeon's knives) give the surgeonfish its name. It erects these weapons hidden in its tail only when it senses danger. Underwater photographer Jeff Rotman learned about the surgical skill of this fish on a night dive in the Red Sea. The flash from his camera startled a surgeonfish sleeping in a cave. (Like many small reef fish, the surgeonfish is shaped like a pancake so it can hide inside narrow openings in the reef.) As it fled the scene, the surgeonfish sliced through Jeff's wetsuit and slashed his thigh. Jeff then had to visit a human surgeon, who gave him six stitches to close the wound.

Lionfish

A close relative of the stonefish, the lionfish does not try to hide the way its cousin does. It practically swaggers. Its bold red or black stripes and aggressive behavior make you think this fish is spoiling for a fight. If a lionfish is threatened, it spreads its side and top fins to warn the intruder that it's not wise to tangle with its venomous spines. A lionfish will often fearlessly charge a much larger challenger. It also spreads its fins to herd smaller fish into a corner of the reef, where it swallows them in a gulp.

A Sea Snake's Bite Can Be Worse Than a Cobra's

Sea Snake

The ocean is home to only a few reptiles, including sea turtles, marine iguanas, and deadly sea snakes. There are several kinds of sea snakes, and the venom of some is stronger than cobra venom. Fortunately, they are usually very timid creatures.

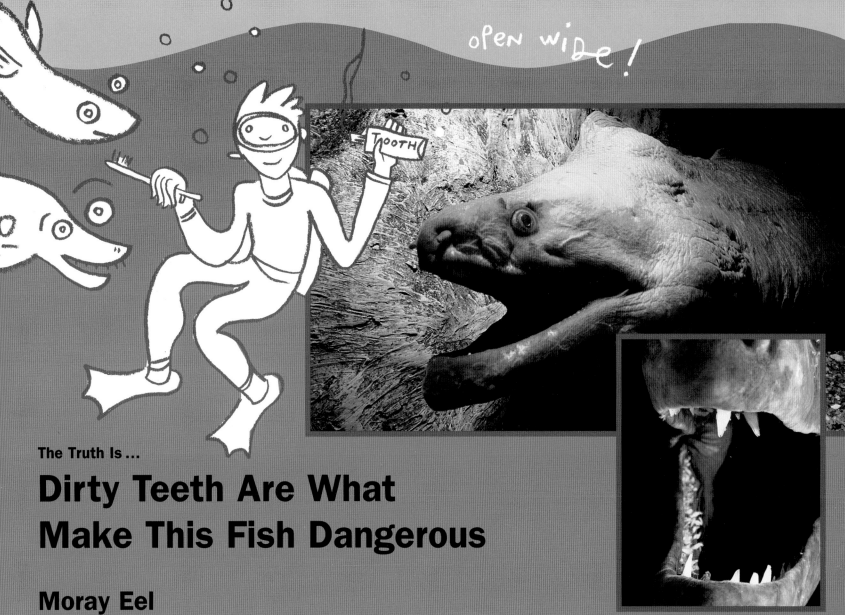

The Truth Is ...

Dirty Teeth Are What Make This Fish Dangerous

Moray Eel

Watching a green moray eel, even from the other side of an aquarium tank, is creepy. It stares at you with its beady little eyes, coiled like a snake ready to strike. It opens and shuts its jaws continually as if it's imagining what you might taste like. (It's really just pumping water over its gills to breathe.) Yellow slime covers its bluish skin, making it look a sickly green. But even though it looks like a snake, it's not. The moray eel is a fish.

A moray eel does not usually attack humans, but it is not forgiving if a diver pokes a hand into its den in the reef. Its needlelike teeth can make deep puncture wounds in your hand. Bacteria on the moray's teeth from partly digested food can then get into the wound and cause an infection.

Biting a Barracuda May Be Worse Than Being Bitten by One

Barracuda

Another fish with infamous teeth is the barracuda. With a body as sleek as an arrow, the barracuda is built for speed. Sometimes called the "tiger of the sea," a barracuda lies in wait until it is ready to ambush its prey. Then it rushes at its target, sometimes cutting it in half with the force of its attack and its razor-sharp teeth.

Divers sometimes claim that they have been stalked by barracuda, which could be true, but most attacks on humans appear to be accidental. Barracuda have been known to attack shiny objects like rings, bracelets, and dive gear or to go after a fish that has been speared by a diver. Neal Watson, an experienced dive master, used to feed a friendly barracuda that hung around his dock. One day Neal was feeding "Charlie" by hand when he was momentarily distracted. Impatient for his dinner, Charlie rushed at the fish and ripped it from Neal's fingers, along with part of Neal's hand. After needing 36 stitches, Neal is no longer "friends" with Charlie.

Barracuda are common in many tropical areas, but they are rarely eaten. Barracuda can carry a poison that is colorless, odorless, and doesn't break down during cooking. This poison, called *ciguatera* (sig-wah-TER-a), causes terrible stomach cramps and vomiting. It is found in 400 tropical fishes, but it is often more concentrated in barracuda, which are at the top of the tropical food chain. After years of searching for the source of this poison, scientists now believe fish get it from eating a species of tiny, floating plants (called *phytoplankton*) that bloom in large numbers at different times, such as after a shipwreck or a hurricane. Scientists are not able to predict when these outbreaks, somewhat like red tides in other parts of the ocean, will occur. All they can say for sure is that animals at the top of the food chain, such as barracuda, can pass along the poison to human diners even though the fish do not get sick from it themselves.

The Truth Is …

Size Matters

The giant Napoleon wrasse is a distant relative of the giant grouper. In the Red Sea, a giant Napoleon wrasse like this one was nicknamed "Moshe" by divers who fed it hardboiled eggs. It would gently slurp the egg from the diver's hand and moments later spit out an empty shell.

Giant Grouper

Imagine being followed for hours by a curious fish that weighs twice as much as you do and is longer than you are. When you watch a giant grouper gulp its prey, you can't help but wonder if this fish could gobble a diver whole. A grouper can distend its mouth—sort of like unhinging its jaws—to enable it to swallow very large prey.

Many giant groupers (also called Goliath groupers) live in caves and ship-wrecks that are also popular with sport divers. They have been known to follow divers for hours, gently nipping at their flippers. A giant grouper reportedly tried to bite the legs of divers working at the base of an offshore oil rig. They escaped inside a diving bell just in time, but not before one diver lost his dive boot to the fish.

Most groupers are friendly and harmless, and divers enjoy bringing them treats. But enough "close-encounter" stories exist that make us wonder if the giant grouper really could swallow a human. If it's true, the victims aren't telling.

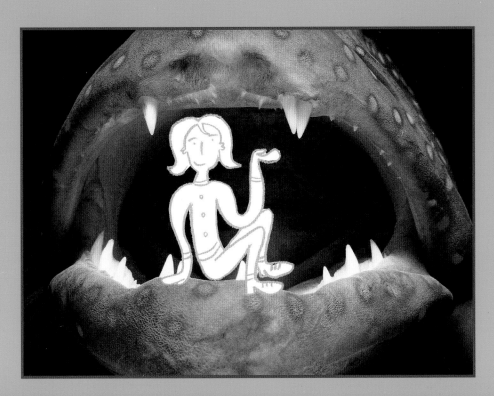

Sharks Are Afraid of This Fish

Puffer

A clumsy puffer paddling through the water doesn't seem like it would be much of a match for a hungry shark. But should the shark be foolish enough to try to eat it, the puffer can inflate itself with water until it becomes a prickly balloon lodged in the shark's throat. Unless the shark can quickly cough it up, both the puffer and the shark will regret the encounter and die a painful death.

For humans, the danger from a puffer also comes from eating it, since some parts of this fish are highly poisonous. Nevertheless, in Japan, it is a delicacy known as "fugu," and it can be prepared only by a licensed fugu chef—who must not only be a good cook but a skilled biologist, surgeon, and artist as well. A fugu chef has to be able to recognize the many different species of puffer fish. Some kinds are always deadly, while others never are. Some are poisonous only at certain times of the year. Many have poison in different parts of their bodies, such as the skin, the liver, and other internal organs. The fugu chef carves out the flesh that is safe to eat and arranges it artistically with vegetables on the plate.

A skilled fugu chef leaves enough poison in the fish to give diners a pleasant tingling sensation around their mouths. A less-talented fugu chef may have no guests at the end of the meal to pay the bill.

OW!

Is It Poisonous or Venomous?

Venomous animals inject poison (also called *toxin* or *venom*) into a victim by stinging, biting, or stabbing. Lionfish and stingrays are venomous fish. Puffers and barracuda are examples of marine animals that are, or can be, poisonous. These fish do their harm after being swallowed.

Danger Lurks under the Sand

Stingray

You could say a ray is really a flattened shark. The stingray is perfectly designed for bottom living. It often buries itself in the sand up to its eyeballs. Its mouth and gills are on its bottom side. Its eyes and breathing holes, called *spiracles*, are located on the top of its head. The stingray takes in water through the spiracles instead of through its mouth, as most fish do. Otherwise it would swallow sand when it pumps water over its gills to breathe. It has heavy, grinding teeth that are designed for crushing the shells of fellow bottom dwellers such as crabs and snails.

A stingray would much rather flee than fight. Its coloring matches the sandy ocean floor, helping to camouflage it. If it is disturbed, it abandons its hiding place and flutters to another spot a few feet away. If you step on it, however, it whips its tail around and slashes your foot with a barbed spine. Sometimes the spine stays in the victim, but the stingray has another spine just behind it to replace the lost one.

There is a shallow bay in Grand Cayman Island known as "Stingray City." Several years ago local fishermen cleaned their catch here and threw the remains overboard. Stingrays began to gather to take advantage of the free meals. Soon tourists began to gather to pet and feed the stingrays. Now the sound of boat engines attracts up to 70 stingrays each day. They swarm from boat to boat to get handouts of squid like eager trick-or-treaters.

Torpedo Ray

Once, photographer Jeff Rotman was swimming across a sandy lagoon trailing his fingers along the bottom. Suddenly he felt an electric shock run up his left arm as his hand brushed the side fin of a torpedo ray buried in the sand. The torpedo ray has two large, kidney-shaped electric organs stacked like batteries under its skin, and they release a shock that can stun prey. Humans usually feel no more than a slight numbing sensation, although one man wading along the shoreline was thrown to the ground by the shock from a torpedo ray. Appropriately, this fish is also known as an electric ray, numbfish, and crampfish.

Stingrays looking for handouts in "Stingray City."

Rays Are Not "Devilfish"

Manta Ray

These gentle, graceful creatures are shaped like kites and "fly" through the water with winglike side fins. They are often very tolerant of people. The gigantic manta ray actually gives rides to divers. Instead of avoiding people, manta rays often approach them.

Some manta rays can grow to be 20 feet (6 meters) across. The manta has been unkindly nicknamed the "devilfish," perhaps because of two fins along the side of its head that resemble devil's horns. These fins sweep schools of fish, shrimp, and plankton into its mouth.

One unexpected trait of these huge animals is their ability to jump clear out of the water. Scientists don't really understand why they do this. Some think it's to get rid of hitchhikers like skin parasites or remoras (shark suckers). This leaping also poses the manta's only real danger to humans, since a ray could accidentally crash into a small boat when it lands.

The Largest Sharks Are Not Dangerous

Basking Shark and Whale Shark

What is as large as a schoolbus and picks up underwater hitchikers? Whale sharks often allow divers to ride on their backs. Basking sharks grow to be 40 feet (12 meters) long, and whales sharks grow to be as much as 50 feet (15 meters) long. That makes each one about as big as a school bus.

These sharks have tiny little teeth that are essentially useless for feeding. But they also have small hooks on their gills, and they swim through the water with their mouths wide open, catching zooplankton (tiny, floating animals), small fish, and krill (small, shrimplike animals) on these hooks. Then they suck in water through their gills to flush the food into their bellies.

Basking Shark

Whale Shark

There Are More Dangerous Sharks Than the Great White

Whenever there is a shark attack on humans, the great white shark gets blamed. But at least two other kinds of sharks, the tiger shark and the bull shark, are often the ones responsible.

Tiger Shark

The tiger shark is the largest shark in most Pacific reefs. It feeds on turtles, stingrays, and smaller sharks. In fact, it will feed on just about anything that comes its way, as proved by the strange objects that have been removed from tiger shark stomachs over the years: license plates, unopened cans of salmon, driftwood, boat cushions, a wallet, a tom-tom drum, and from time to time, human body parts. Tiger sharks have clearly earned their nickname "garbage can sharks."

Bull Shark

Bull sharks are known as "freshwater sharks" because
they have been found in rivers and lakes thousands of miles
from the ocean, including Lake Nicaragua, the Zambesi
River, and the Mississippi River. They are considered
dangerous to humans wherever they are found. Perhaps bull
sharks leave the ocean to swim up rivers in search of food or
to get rid of skin parasites that can't survive in freshwater.
Like tiger sharks, they will eat other sharks and rays and
almost anything else. In rivers, they have been known to
attack young hippos.

Great White Shark

Broad and bulky, the great white shark could be considered the football player of sharks. It is not sleek and streamlined like some other sharks. Although it can still move fast when it needs to, it relies on strategy and power to take down its opponent. A great white stalks its prey from 100 feet (30 meters) below and then surges up to take a 20-pound (9-kilogram) bite out of its victim. Then the shark waits until the injured animal weakens and stops struggling before it begins feeding. It prefers fatty food like seals, sea lions, and whales. Compared to those blubbery prey, a bony human might not be very appetizing to a great white shark.

The hammerhead shark is considered a "man-eater."

The nurse shark is a placid bottom-feeder, harmless to humans.

As long as people have been sailing the seas, they have been looking for ways to prevent shark attacks. In the 1700s French fishermen working in the Mediterranean Sea greatly feared that sharks would attack and swallow their small fishing vessels. If any hungry shark followed their boat, they threw loaves of bread to it. If that didn't satisfy it, the fishermen would lower a sailor on a rope to the water's surface to make scary faces at the shark. No

Helpful Hints

What is the best thing to do should you find yourself in the company of a shark? Dr. Eugenie Clark, one of the world's experts on sharks, advises simply, "Be polite." That advice works equally well for all the dangerous animals in this book. If you don't hurt them (by stabbing, poking, grabbing, or kicking them), they won't hurt you (by biting, stabbing, shocking, or poisoning you). It's important to remember that you are a guest in their home. Underwater, good manners could literally save your life.

Despite your best efforts, you still might get injured while exploring the ocean. Here are some suggestions for making the best of the situation.

- **Call a doctor** or poison control center.

- **Wash any wound thoroughly** and check to make sure no bits of coral or venom remain in the cut.

- **Don't be too quick to get back into the water** if you are stung and yet feel okay. Sometimes, severe reactions don't develop until several hours later. Also, having been stung once may make you more sensitive to the venom if you are stung again.

Household remedies sometimes help to treat jelly stings. Different sources recommend various solutions to neutralize the stinging cells, such as vinegar, rubbing alcohol, sugar, soap, lemon juice, and shaving cream. Don't wash the skin with freshwater because this can cause dried stinging cells to fire.

Prevention is best:

- Wear sneakers or watershoes when wading in the water.
- Shuffle your feet to scare up stingrays.
- Don't swim at dawn, dusk, or night, when big predators are hunting.
- Don't swim in murky water, where fish might mistake you for something good to eat.

Glossary

camouflage
any kind of coloring that helps an animal blend in with its surroundings

ciguatera (sig-wah-TER-a)
a poison found in many tropical fish that is caused by a kind of toxic phytoplankton

ecosystem
how living things and their environment function as a unit

food chain
a linear sequence of who eats whom (such as plant plankton, animal plankton, shrimp, fish, human)

fugu
a Japanese delicacy made from puffer

invertebrate
an animal without a backbone. More than 90 percent of all the animals in the ocean are invertebrates, including sea stars, lobsters, snails, octopuses, jellies, and more

jelly or jellyfish
an animal with a gelatinous, umbrella-shaped body and tentacles that can sting and capture prey

kraken
a mythical giant sea monster with many arms that could sink a ship

krill
shrimplike animals up to 2 inches (5 centimeters) long, which are an important part of the marine food chain

phytoplankton
plant plankton

plankton
tiny drifting plants or animals, including microscopic algae and the baby stages of many sea animals. The word plankton literally means "wanderer"

predator
an animal that eats other animals

prey
an animal that is eaten

spiracle
a small opening on either side of a ray's eyes where water enters. Spiracles connect to the gills, which take oxygen from the seawater

symbiosis
literally, "living together." Two different species live together in a relationship in which both benefit; for example, a sea anemone and a clownfish

vertebrate
any animal with a backbone and an internal skeleton, such as fish, lizards, and humans

zooplankton
animal plankton that feed on phytoplankton or other zooplankton

Bibliography

Dive Deeper—but with Caution

There are a number of books that discuss the individual animals mentioned here in more detail. Whenever you read these and other books, ask yourself, "Is the author being fair and objective?" Especially when writing about animals that could be dangerous to people, it is tempting to describe their behavior as "good" or "bad." Animals in the ocean do what they do in order to survive. For them, "good" is living another day. "Bad" is becoming someone else's meal!

Books for young readers

The Truth about Great White Sharks by Mary M. Cerullo. San Francisco: Chronicle Books, 2000.

The Octopus: Phantom of the Sea by Mary M. Cerullo. New York: Cobblehill, 1997.

Coral Reef: A City That Never Sleeps by Mary M. Cerullo. New York: Cobblehill, 1996.

Beneath Blue Waters: Meeting with Remarkable Deep-Sea Creatures by Deborah Kovacs and Kate Madin. New York: Viking, 1996.

Websites

http://www.mbayaq.org Monterey Bay Aquarium, Monterey, California

http://www.neaq.org New England Aquarium, Boston, Massachusetts

http://www.aqua.org National Aquarium in Baltimore, Baltimore, Maryland

http://www.sheddnet.org Shedd Aquarium, Chicago, Illinois

http://www.coralreef.noaa.gov Coral Reefs, National Oceanic and Atmospheric Administration

Index